YOUR KNOWLEDGE HAS VALUE

- We will publish your bachelor's and master's thesis, essays and papers

- Your own eBook and book - sold worldwide in all relevant shops

- Earn money with each sale

Upload your text at www.GRIN.com and publish for free

Blessing Adegoke

Assessment of criteria for selection of timber as building material for building construction projects within Lagos Metropolis

GRIN Publishing

Imprint:

Copyright © 2009 GRIN Verlag, Open Publishing GmbH
Print and binding: Books on Demand GmbH, Norderstedt Germany
ISBN: 978-3-656-12680-5

This book at GRIN:

http://www.grin.com/en/e-book/188780/assessment-of-criteria-for-selection-of-
timber-as-building-material-for

ASSESSMENT OF CRITERIA FOR SELECTION OF TIMBER AS BUILDING MATERIAL FOR BUILDING CONSTRUCTION PROJECTS WITHIN LAGOS METROPOLIS

CHAPTER ONE

1.0   INTRODUCTION

1.1   Background to the study

Shelter is one of the basic needs of people in every society. Man strives to have decent homes and environment; this is applicable in both rural and urban communities by the use of good building materials. The earliest forms of houses were built by mud which can still be found in some areas in the country. Gradually, development and structural advancement began to take place in the world of building. As man moved from Stone Age to computer age; this was further enhanced by the growth and advent of new building materials and modern technology. As a result of changes in design, environmental decoration and taste of individuals, the old and crude methods of building productions were rendered obsolete. Hence, there is need for a holistic approach for integrating sustainability principles into material selection decision making at the design stage of building project (Akadiri, 2011).

Careful selection of environmentally sustainable building materials is the easiest way for architects and building designers to begin incorporating sustainable design principles in buildings. Traditionally, price has been the foremost consideration when comparing similar materials or materials designated for the same function. However, the "off-the-shelf" price of a building component represents only the manufacturing and transportation costs, not social or environmental costs (Jong-Jin Kim, 1998).

Rapid population growth and the continuous growth of industrialization throughout the world together with increasing living standards have turned the creation of the built environment into a rising threat to the natural environment. Building and the environment are inextricably linked (Akadiri, 2011). Buildings have a significant and continuously increasing impact on the environment and they use a considerable number of material resources; buildings accounts for one-sixth of the world's fresh water withdrawals, one-quarter of its wood harvest, and two-fifths of its materials and energy flows. The extraction of building materials such as wood, steel and cement causes water, air pollution, and degrades the land on which people live and play (ICEST 2011).

An important task in the implementation of objectives at the design development stage of a building project is the selection of sustainable building materials to be used in the building project. At the same time, building materials are the substance of economic life in the industrial sector. It acquires concern on how to select the right building materials, because the choice of materials is likely to affect the building's energy need for the material (ICEST 2011).

The idea behind assessment of criteria for selecting building materials is a multidisciplinary process that engages every one from government and utilities through to contractors and operators as well as investors at the very preliminary stages of a new build or retrofit project. Communities benefits as a result of producing modern attractive structures that defines the standard of the community at a point in time.

It is suggested that increasing the emphasis on wood as building material could have significant implications for global energy requirements and global carbon dioxide emission unlike concrete and steel. The realization of the importance of material selection in construction of building and other related structures makes it necessary to look for a way of merging the respective qualities of the various building materials with the construction needs in order to achieve a reliable and quality building structures.

1.2   STATEMENT OF RESEARCH PROBLEM

As a result of great urge for development, the construction industry in Nigeria is rapidly developing in handling simple to complex building construction projects; there have been observations from various sites (within Lagos State), where construction activities are going on, that selection of building materials are not properly enhanced to their maximum potentials. As a result of this, the study of assessment of criteria for selection of building materials on project quality becomes necessary.

However, the sustainability of a building depends on the decisions taken by a number of stakeholders in the construction process: owners, managers, designers, firms, contractors, engineers, construction industries and investors. The pace of actions towards sustainable application depends on the awareness, knowledge as well as an understanding of the consequences of individual actions. Among these is the environmentally responsible approach to the selection of building materials (Anderson, 2009).

The selection of building materials is one of several factors that can impact the sustainability of a project; an appropriate choice of material selection in a design process plays an important role during the life cycle of a building (Zhou, 2008). Understanding the environmental issues surrounding the extraction of raw materials, the manufacture of construction materials, and their effects in use, is important to ensure sustainability and maximize the potentials of these materials. Structural designers identify selection of sustainable building materials as the basis for incorporating sustainable principles in building projects (Godfaurd, 2005).

1.3   RESEARCH QUESTIONS

The following are the research questions as related to this study:
1. What are the criteria for selecting sustainable building materials?
2. What are the effects of decision-making process in selecting building materials?
3. What are the factors to be considered in the selection of building materials?

1.4   AIM OF THE STUDY

The aim of this study is to assess the criteria for selection of building materials for various projects.

1.5   OBJECTIVES OF THE STUDY

The objectives of the study are:
1. To identify significant criteria for selecting sustainable building materials.
2. To assess the decision-making process in the selection of building materials.
3. To examine the factors to be considered in selection of building materials.

1.6   HYPOTHESIS OF THE STUDY

H0:   Assessing the criteria for selecting building materials improves the overall performance of the building.

H1:   Assessing the criteria for selecting building materials has no significant effect on the overall performance of the building.

1.7   Significance of the study

In view of the fact that construction in the building industry is a dynamic process, the assessment of criteria for selection of building materials will assist the various participants of building projects within the construction industry in terms of the likely effects and consequences of materials selection and supply methods. The study will provide information and reveal various constraints to be considered and how to proffer solutions to them.

It is very important to point out that despite the fact that a lot of challenges are involved in the material selection for any construction project; it becomes more effective in terms of time, cost and quality. Also, material selection in the construction industry is considered right from the inception to completion of a project. According to Akintoye, (2001), Projects are not devoid of challenges of selecting appropriate materials. Hence this research was effectively carried out so as to let the professionals in the construction industry be more sensitive to the criteria of material selection for any construction projects and how best to manage them as possible in other to achieve the required structure. Thus, this study is of importance to both the owner and the end users.

The importance of selection of building materials to the environment at large cannot be over-emphasized. Analysis of building production, from the gathering of raw materials to their ultimate disposal, provides a better understanding of the long-term costs of materials. These excess cost can be critically control by the principles of Life Cycle Analysis which provides important guidelines for the selection of building materials.

A material's life cycle can be organized into three phases: Pre-Building; Building; and Post-Building. These stages parallel the life cycle phases of the building itself. Academicians critically evaluate building materials' environmental impact at each stage and access the cost-benefit analysis over the lifetime of a building, rather than simply an accounting of initial construction costs (Jong-Jin Kim, 1998).

1.8   Scope and limitation of the study

Building construction focus is on minimizing the initial building cost, achieving a quality standard of building product within a reasonable period of time. It has, however become obvious that it is unfavorable to base the choice between material alternatives solely on the initial cost alone (Kishk, 2003). Hence, the scope of this research will be restricted to building construction projects that are carried out within Lagos State. The research study will be centered on evaluating the selection of "timber" as a building material.

1.9   Definition of terms

The following are the definition of terms used in the research;

Buildability: The extent to which the design of a building facilitates ease of construction subject to the overall requirements for the completed building (constructability).

Building: Construction works that has the provision of shelter for its occupants or contents as one of its main purposes and is usually enclosed and designed to stand permanently in one place.

Building cost: The total cost of all Building Elements for a particular Building.

Building element: The Building Element comprises all elements that are primarily part of the construction of a building, i.e. it's structural and space separating system.

Building Project: Is a temporary endeavor, having a defined beginning and end (usually constrained by date, but can be by funding or deliverables) undertaking to meet unique goals and objectives.

Criteria: Is a characterizing standard or trait on which a judgment or decision may be based.

Life-Cycle Analysis / Assessment: Compilation and evaluation of the inputs, outputs, and the potential environmental impacts of a product system throughout its life cycle.

Material: A primary substance or by-product (e.g. sand or slag) used to create a larger entity. Materials are often referred to in specifications as products.

CHAPTER TWO

2.0 LITERATURE REVIEW

2.1 Introduction

As the earth population continues to explode and developing nations begins to use their share of the world's resources, it is necessary to ascertain how we, as a planet use our earth's precious resources. During construction or at the end of useful building life, construction materials and components are often discarded with construction debris accounting for a large percentage of landfill waste. Inappropriate use of building materials that emits chemicals can pollute the indoor environment contributing to poor indoor air quality (IAQ) with some new building chemical concentrations (Lynn, 1999).

Manufacture, construction and use of buildings and building materials make a significant environmental impact internally, locally and globally. But it is not easy to deliver information to make adequate holistic decisions considering the whole life cycle of building. Decisions in sustainable building integrate a number of strategies during the design, construction and operation of building projects. Selection of sustainable building materials represents an important strategy in the design of a building (Seongwon, 2002).

Managing Authorities often express commitments to sustainability, yet the assessment of sustainability in large building projects remains limited to traditional environmental assessment method, leaving the energy related aspects aside. Furthermore, while traditional method focuses on mitigating negative effects, achieving true sustainable building materials demands that each new undertaking make a positive contribution to desirable and durable futures. This can only be achieved if Managing Authorities are able to consciously and publicly specify and use sustainability-centered criteria to identify good and sustainable building materials for various construction projects and to weigh trade-offs. A sustainability assessment should cover all aspects of the core requirements of building material selection including the creation of basic criteria, handling trade-offs, practicalities in application, implications for process design and uses in decision-making, as well as examining the range of tools and examples available to assist implementation of sustainability assessment (Donkelaar, 2010).

In this assessment we will therefore not only look at mitigating negative environmental impacts, but also at aspects of social and economic development, as sustainability assessment of building material selection is based on environmental, economic and social criteria.

The World Health Organization (WHO) estimates approximately 30 percent of all buildings will have indoor air quality (IAQ) concerns during the facility occupancy. Green or environmentally friendly materials can help create more sustainable, healthful, and ecologically sensitive buildings. This is achieved through environmental material assessment and green building specifications (Lynn, 1999).

2.2   Environmental Material Assessment

The assessment of environmental materials begins with establishing criteria for evaluating building materials. These criteria should compliment the overall environmental project goals. Based on the environmental material criteria established for a green building project, selection of appropriate building products, and systems can be accomplished.

The environmental material criteria may vary per project. Criteria may also vary depending on whether a project is new construction, a renovation of an existing building, or whether site work is associated with the project.

With heightened awareness of environmental pollution, natural resource depletion and accompanying social problems, sustainable development and sustainable construction have become a growing concern throughout the world (ICEST 2011). Buildings are one of the heaviest consumers of natural resources and account for a significant portion of the greenhouse gas emissions. In many developing countries, rapid economic growth is outstripping infrastructure supply, the use of prefabrication offers significant advantages, yet appropriate criteria for applicability assessments of materials to a given building project have been found to be deficient. Decisions to use prefabrication are still largely based on cost-based evaluation when comparing various construction methods and construction materials. Holistic criteria are needed to assist with the selection of appropriate construction materials in buildings construction during early project stages. (Chen, 2010).

A material's life cycle can be organized into three stages: Pre-Building; Building; and Post-Building. These stages parallel the life cycle phases of the building itself, the evaluation of building materials' environmental impact at each stage allows for a cost-benefit analysis over the lifetime of a building, rather than simply an accounting of initial construction costs.

2.2.1 Three Phases of Building Materials

These three life-cycle phases relate to the flow of materials through the life of a building (Jong-Jin Kim, 1998). They are:

2.2.1.1 Pre-Building Phase

The Pre-Building Phase describes the production and delivery process of a material up to, but not including, the point of installation. This includes discovering raw materials in nature as well as extracting, manufacturing, packaging, and transportation to a building site. This phase has the most potential for causing environmental damage. Understanding the environmental impacts in the pre-building phase will lead to the wise selection of building materials. Raw material procurement methods, the manufacturing process itself, and the distance from the manufacturing location to the building site all have environmental consequences. An awareness of the origins of building materials is crucial to an understanding of their collective environmental impact when expressed in the form of a building.

2.2.1.2 Building Phase

The Building Phase refers to a building material's useful life. This phase begins at the point of the material's assembly into a structure, includes the maintenance and repair of the material, and extends throughout the life of the material within or as part of the building.

Construction: The material waste generated on a building construction site can be considerable. The selection of building materials for reduced construction waste, and waste that can be recycled is critical in this phase of the building life cycle.

Use/Maintenance: Long-term exposure to certain building materials may be hazardous to the health of a building's occupants. Even with a growing awareness of the environmental health issues concerning exposure to certain products, there is little emphasis in practice or schools on choosing materials based on their potential for outgassing hazardous chemicals, requiring frequent maintenance with such chemicals, or requiring frequent replacements that perpetuate the exposure cycle.

2.2.1.3 Post-Building Phase

The Post-Building Phase refers to the building materials when their usefulness in a building has expired. At this point, a material may be reused in its entirety, have its components recycled back into other products, or be discarded. From the perspective of the designer, perhaps the least considered and least understood phase of the building life cycle occurs when the building or material's useful life has been exhausted. The demolition of buildings and disposal of the resulting waste has a high environmental cost. Degradable materials may produce toxic waste, alone or in combination with other materials. Inert materials consume increasingly scarce landfill space. The adaptive reuse of an existing structure conserves the energy that went into its materials and construction. The energy embodied in the construction of the building itself and the production of these materials will be wasted if these "resources" are not properly utilized.

2.2.2 Three basic steps of product selection

Product selection can begin after the establishment of project-specific environmental goals. The environmental assessment process for building products involves three basic steps.(Froeschle, 1999)

- Research
- Evaluation
- Selection

2.2.2.1 Research

This step involves gathering all technical information to be evaluated, including manufacturers' information such as Material Safety Data Sheets (MSDS), Indoor Air Quality (IAQ) test data, product warranties, source material characteristics, recycled content data, environmental statements, and durability information. In addition, this step may involve researching other environmental issues, building codes, government regulations, building industry articles, model green building product specifications, and other sources of product data. Research helps identify the full range of the project's building material options.

2.2.2.2 Evaluation

This step involves confirmation of the technical information, as well as filling in information gaps. For example, the evaluator may request product certifications from manufacturers to help sort out possible exaggerated environmental product claims. Evaluation and assessment is relatively simple when comparing similar types of building materials using the environmental criteria. For example, a recycled content assessment between various manufacturers of medium density fiberboard is a relatively straightforward "apples to apples" comparison. However, the evaluation process is more complex when comparing different products with the same function. Then it may become necessary to process both descriptive and quantitative forms of data.

A life cycle assessment (LCA) is an evaluation of the relative "greenness" of building materials and products. LCA addresses the impacts of a product through all of its life stages. Although rather simple in principle, this approach has been difficult and expensive in actual practice.

2.2.2.3 Selectio

This step often involves the use of an evaluation matrix for scoring the project-specific environmental criteria. The total score of each product evaluation will indicate the product with the highest environmental attributes. Individual criteria included in the rating system can be weighted to accommodate project-specific goals and objectives.

Some building materials may be chosen because of their adaptability to new uses. Steel stud framing, for example, is easily reused in interior wall framing if the building occupants' needs should change and interior partitions need to be redesigned (modular office systems are also popular for this reason). Ceiling and floor systems that provide easy access to electrical and mechanical systems make adapting buildings for new uses quick and cost-effective (Jong-Jin Kim, 1998).

2.3   Timber as a Building Material

Timber is one of man's oldest materials. It is a renewable, naturally occurring organic polymer, unique in a world of synthetic and composite building materials. As a building material, timber is valued for its character, warmth and texture. The use of timber as construction material is a chance to save the world's forests as it is directly linked to forest conservation and the planting of new trees to meet the green environmentally friendly requirements.

Timber is a versatile, strong and adaptable material both in its raw form as wood and as a range of products. Timber has been used in construction throughout history primarily due to its relative strength, ease of use and fixing and the fact it is a readily available natural resource. Although the ease of fixing and modifying timber components during construction has proved popular with builders, for instance notching and drilling joists for services, these types of modification turn a generic joist of uniform section into a joist that is tailored specifically for the building it is installed in. The wide distribution of timber, its ready availability, variety of uses and relative ease of handling and conversion, have all contributed to the wide acceptance of timber in the building industry. Timber also has the advantage of being low in the consumption of energy needed in its processing, and hence in its overall impact on the biosphere (Hobbs, 2008).

The construction of buildings involves incorporation of many materials which includes marble, stone, glass, steel rods etc and of course timber. Timber is a primary building material which is relatively cheap. Timber is the only building material that, when properly used, positively benefits the planet and our relationship to it. Timber is a renewable resource, produced by solar energy; it's also a processed wood. Timber has roughly the same structural properties as steel or reinforced concrete, yet producing a cold rolled steel section equivalent to a 50mm x 300mm solid untreated softwood section of the same stiffness requires nineteen times much energy. Structural hardwood would show an even greater difference because a smaller section would have the same stiffness. Even in large-span structures, timber construction requires less than a quarter of the energy of concrete (Hurley, 2009).

2.3.1 Structural Properties of Timber

The properties of timber are determined majorly by the structure of the wood i.e. the type of cells and the chemical composition of the cell walls and cell contents, and by the proportion of the various cell types present in the wood. The distribution and arrangement of these wood elements vary considerably in different species of trees.

2.3.1.1 Strength and Stiffness

Strength and stiffness are allied to density and as density of boards from an individual species varies considerably, so does the strength. One aspect of the strength is the load carrying capacity of a length of timber, usually expressed in "modulus of rupture in static bending". Modulus means measures, so it's a measure of the maximum load-carrying capacity when that load is applied slowly at the centre of the beam. Stiffness describes a length of timber's resistance to deflection under load and 'modulus of elasticity in static bending' is the measure.

2.3.1.2 Increasing Bendability

Plasticization (softening) of timber can be achieved chemically by microwave radiation, by streaming or by boiling. If pieces of wood of smaller cross section are soaked in a stream bath or near-boiling water for sufficient time for the wood to reach an internal temperature of 90-95 degrees-Celsius and moisture content of 18-20%, the piece becomes ductile and thus able to be bent into a permanent shape without fracturing. If then clamped in a shaped form of the desired curve, the piece will maintain somewhere near that bend after cooling and drying out.

2.3.1.3　　　Moisture Content and Density

The moisture content of timber is the quantity of moisture contained in it, expressed as the percentage of the dry weight. In green timber, moisture is contained within the cells and the cell walls. Timber as felled has considerable moisture content present as 'free moisture' within the cell cavities and bound moisture saturating the cell walls. Freshly sawn lumber will lose perhaps 50% of its total weight, shrinks somewhat and becomes much stronger, harder and more durable during the seasoning (drying and stabilizing) process. The seasoning process also improves timber workability and the bonding adhesives and surface finishes.

$$\text{Moisture Content} = \frac{\text{Weight of Water}}{\text{Weight of totally dry wood}}$$

The density of timber is an indication of its strength and stiffness.

2.4　Criteria for Building Materials Selection

The most commonly quoted sustainability criteria for selection of building materials is that they should score well in most or all of the following areas (Buchanan, 2002). Materials should be:

1. Resource efficiency
2. Indoor air quality
3. Energy efficiency
4. Water conservation
5. Affordability
6. Non-polluting
7. Minimum waste

2.4.1　Resource Efficiency

Can be accomplished by utilizing materials that meet the following criteria:

1. ***Recycled Content***: Products with identifiable recycled content, including postindustrial content with a preference for postconsumer content.

2. ***Natural, plentiful or renewable***: Materials harvested from sustainably managed sources and preferably have an independent certification (e.g., certified wood) and are certified by an independent third party.

3. ***Resource efficient manufacturing process***: Products manufactured with resource-efficient processes including reducing energy consumption, minimizing waste (recycled, recyclable and or source reduced product packaging), and reducing greenhouse gases.

4. ***Locally available***: Building materials, components, and systems found locally or regionally saving energy and resources in transportation to the project site.

5. ***Salvaged, refurbished, or remanufactured***: Includes saving a material from disposal and renovating, repairing, restoring, or generally improving the appearance, performance, quality, functionality, or value of a product.

6. ***Reusable or recyclable***: Select materials that can be easily dismantled and reused or recycled at the end of their useful life.

7. ***Recycled or recyclable product packaging***: Products enclosed in recycled content or recyclable packaging.

8. ***Durable***: Materials that are longer lasting or are comparable to conventional products with long life expectancies.

2.4.2 Indoor Air Quality (IAQ)

Is enhanced by utilizing materials that meet the following criteria:

1. ***Low or non-toxic***: Materials that emit few or no carcinogens, reproductive toxicants, or irritants as demonstrated by the manufacturer through appropriate testing.

2. ***Minimal chemical emissions***: Products that have minimal emissions of Volatile Organic Compounds (VOCs). Products that also maximize resource and energy efficiency while reducing chemical emissions.

3. ***Low-VOC assembly***: Materials installed with minimal VOC-producing compounds, or no-VOC mechanical attachment methods and minimal hazards.

4. ***Moisture resistant***: Products and systems that resist moisture or inhibit the growth of biological contaminants in buildings.

5. ***Healthfully maintained***: Materials, components, and systems that require only simple, non-toxic, or low-VOC methods of cleaning.

6. ***Systems or equipment***: Products that promote healthy IAQ by identifying indoor air pollutants or enhancing the air quality.

2.4.3 Energy Efficiency

Is an important feature in making a building material environmentally sustainable. Depending on type, the energy efficiency of building materials can be measured with factors such as R-value, shading coefficient, luminous efficiency, or fuel efficiency. The ultimate goal in using energy efficient materials is to reduce the amount of artificially generated power that must be brought to a building site. Materials, components, and systems that help reduce energy consumption in buildings and facilities.

2.4.4 Water Conservation

Can be obtained by utilizing materials and systems that meet the following criteria:

Products and systems that help reduce water consumption in buildings and conserve water in landscaped areas.

2.4.5 Affordability

Can be considered when building product life-cycle costs are comparable to conventional materials or as a whole, are within a project-defined percentage of the overall budget.

2.4.6 Non-polluting

Are less hazardous to construction workers and building occupants. Many materials adversely affect indoor air quality and expose occupants to health hazards. Some materials, like adhesives, emit dangerous fumes for only a short time during and after installation; others can reduce air quality throughout a building's life.

2.4.7 Minimum wastes

Many building materials come in standard sizes, based on the 4' x 8' module defined by a sheet of plywood. Designing a building with these standard sizes in mind can greatly reduce the waste material created during the installation process. Efficient use of materials is a fundamental principle of sustainability. Materials that are easily installed with common tools also reduce overall waste from trimming and fitting.

Wood as a building material

Wood as one of the most commonly used material for building construction fits well with most of the above criteria. Wood is a renewable resource, containing a large quantity of solar energy, hence requiring low amounts of fossil fuel for manufacturing, and providing an independent source of energy when waste wood is burned. Using waste wood for energy production results in reduced use of fossil fuel, compared with recycling of metals and glass which use additional fossil fuel energy. Energy from waste wood is solar energy which has been stored in the wood for a few years.

Wood is often sourced locally, although is sometimes transported around the world, so the energy implications of long distance shipping need to be investigated. Wood can be recycled, but not in the extensive manner of meltable materials like metals and glass. Modern timber construction produces little waste, but that produced can be burned for energy. The production of wood is generally non-polluting at all stages although there have been problems in the past with polluted sites from chemical preservative processes. There have been many international studies investigating the amount of energy required to manufacture building materials. Some of that energy comes from hydro or geothermal electricity, and some from burning of fossil fuel, with very little from burning of wood waste. Burning of fossil fuels is associated with increasing levels of carbon dioxide (CO_2) in the atmosphere. It is possible to calculate the fossil fuel component of energy used to make building materials, and the building industry is searching for ways to reduce such emissions (Buchanan, 2002).

Temperature and moisture conditions are, in general, the two major factors influencing the long-term performance of elements made of porous mineral building materials. Degradation of components is accelerated by temperature and moisture induced stresses which can lead to cracks and in turn a surface more vulnerable to other degradation agents. The degradation rate depends on both the environmental conditions and the material-inherent and component design properties. Extreme and rapid temperature fluctuations as well as moderate diurnal and seasonal temperature cycles cause thermal stresses and strains in some materials, resulting in expansion or contraction and eventual deformation such as cracking or fracture. Material properties such as thermal expansion, elasticity and tensile strength determine if cracking occurs either immediately when the surface temperature drops below the initial temperature after rapid cooling or after a period of time if alternating or repeated stresses result in creep and fatigue (Kus, 2005).

2.5   Construction Materials and Components

The building and construction sector should, apart from technical and financial requirements, be based on the information and knowledge being generated by a strive for, or requirements on, sustainable construction. This includes all aspects from an environmentally conscious design and construction process, with pertinent choice of materials and products, to a healthy indoor environment.

Performance requirements and expectations on a system level, with the objective of sustainability, healthiness and environmental friendliness in building and construction, should be fed back on the property profiles of materials and components during all stages of the life cycle of the building. This view could be denoted as Materials and Systems Engineering.

Materials and components are affected by the ambient in-use environment, and require resources during their whole life cycle from raw-material recovery and production, manufacturing, use, maintenance, recycling/re-use, to rest-material handling. All these activities have environmental, economic and social consequences. These consequences, in turn, feed back to the materials, their in use environment and future resource needs (Sjostrom, 2002).

2.6   Materials Selection

'Green' concept is emerging in Asian sub-continent but needs to be made sustainable to cover broad spectrum of Sustainability (Lynn, 1999). Hence, it is high time that countries develop their building material assessment tools to cover all areas like: Environmental, social, economical, safety and cultural. Sustainability assessment methods in material selection for building projects have a major role to play in introducing sustainability values and principles into mainstream construction practice. As noted by Donkelaar, (2010), building environmental assessment methods are doing large contribution for higher environmental expectations and are directly and indirectly influencing the performance of buildings. The development of new tools for Sustainable buildings is shifting from developed countries to developing countries (Bhatt, 2010).

Generally, selection of material for a specific building project is one of the most important decisions in the construction industry. Expert judgments required that, the materials should be suitable with respect to the assessment of the cost of production of such structure and to meet the material selection criteria and service performance specifications. Such consideration are not mutually exclusive but all decisions about them must be recognized as compromises, balancing trade-off between different levels of reliability, performances and maintainability in other to achieve minimum lifetime cost for the building.

Traditionally, the focus in construction is on minimizing the initial building cost. It has, however, since the 1930s become obvious that it is unfavorable to base the choice between material alternatives solely on the initial cost alone (Lynn, 1999). An inefficient building imposes a cost penalty on the client throughout its lifetime. While the client has an incentive to minimize whole life costs, the contractors and consultants do not, as they have no long-term interest in the building and are not accountable for performance in use.

2.7   Factors to Be Considered During Material selection

It is very important to select suitable and optimal materials for dwelling or public purpose buildings, a decision was needed for designing alternatives means to evaluate building structures for both qualitative and quantities attributes as contained in the process. The following are recommended environmental material selection criteria for use in a building system assessment and evaluation (Lynn, 1999):

1. **Low Toxicity:** Materials the manufacturers demonstrates to have reduced toxicity or are non-toxic and avoid carcinogenic compounds and ingredients.

2. **Minimal Emission:** Products that have minimal chemical emissions, emits low or no volatile organic compounds (VOCs), and avoid the use of chlorofluorocarbons (CFCs).

3. **Low Volatile Organic Compounds (VOC) Assembly:** Materials installed with minimal VOC, producing compounds or no VOC mechanical attachment methods and minimal hazards.

4. **Recycled Content:** Products with identifiable recycled content in the materials including post-industrial content with a preference for post-consumer content.

5. **Resource Efficient:** Products manufactured with resource-efficient process including reduced energy consumption, minimizing waste, and reducing green house gases.

6. **Recyclable:** Material that is recyclable at the end of their useful life.

7. **Reusable:** Building components that can be reused or salvaged.

8. **Sustainable:** Renewable natural materials harvested from sustainably managed sources and preferably that have an independent certification.

9. **Durable:** Materials that are longer lasting and are comparable to conventional products with long life expectancies.

10. **Moisture:** Products and systems that resists moisture or inhibit the growth of biological contaminants in buildings.

11. **Energy Efficient:** Materials, components and systems that help reduce energy consumption in buildings and facilities.

12. **Water Conservation:** Products and systems that help reduced water consumption in buildings and conserve water in landscaped areas.

13. **Healthfully Maintained:** Materials, components, or systems that requires only simple, non-toxic or low-VOC methods of cleaning.

14. **Local Products:** Building materials, components, and systems found locally or regionally saving energy and resources in transportation to the project site.

15. **Affordable:** Building product life-cycle cost comparable to conventional materials or as a whole is within a project-defined percentage of the overall budget.

2.8   Life Cycle Assessment (LCA) Principles

In order for any team to be able to make effective decisions about green building strategies, they must be able to demonstrate the use of Life Cycle Assessment (LCA) principles. This concept takes a given building material or system and compares it to other alternative building materials and systems – from the perspective of the entire life of the respective material or system. This analysis considers all environmental aspects of the respective technology – from the origin of the raw materials that are extruded from the earth – to transportation of those materials to the manufacturing plant or mill, and on throughout the entire manufacturing and production and delivery process. LCA takes into consideration all aspects of the production of that end product – all embodied energy – including transit of all materials and the end product, all energy used in extraction and production, all water used, and all pollutants and resulting waste products that are generated and released into the air, water and land as a result of the production of that material.

In addition, LCA takes into consideration the installation of the product and the long term durability, operation and maintenance of the material. End use and disposal, re-use, and/or the energy and water and resources associated with recycling that material (compared to another material) is also factored into a true LCA analysis. A material that has a low embodied energy and low impact on the environment while also achieving attributes of durability and having a long life will be favored over one with a high embodied energy or high use of water and/or one which is less durable and has a shorter life span and is not easily recyclable or re-usable. In true LCA, an emphasis is placed on reuse and recycling vs. down-cycling of materials and resources. LCA is perhaps one of the most sophisticated aspects of green building design as many of these attributes are hard to quantify.

2.9   Energy

Energy use in buildings is an aspect of green design that has perhaps more profound long term environmental impact than any other single attribute. In fact – while much time and analysis (LCA) can be invested comparing one building finish or material with another in terms of embodied energy or resulting pollution and adverse environmental attributes, the bottom line is that the actual energy use of the lighting, electrical and heating, ventilation and cooling systems in a building will have far greater implications in terms of energy use and pollution – over the life of the building – than will all of the building materials used in the building multiplied 100's of times over. Design decisions that impact the buildings operation and energy use are the most profound from an environmental aspect.

As regards the properties of non-load bearing parts of buildings and hence building materials and products, the state-of-the-art is still more complex, but the performance road is as evident. For instance, energy conservation and efficiency of building is normally expressed as a performance requirement which then feeds back on the used materials and components. The requirements have to be declared, assessed and validated for each product. The stochastic variability of loads, e.g. degradation agents, and of the material or product properties should ideally be taken into account in a performance assessment. The objective is to in general develop the tools used in engineering design of load bearing and non-load bearing building parts towards the same conceptual approach (Sjostrom, 2002).

By becoming aware of which building materials and elements of a building have the lowest environmental impact, architects or building designers can encourage the marketing of sustainable buildings by specifying the more environmentally friendly products and redesigning buildings to reduce the largest element contributors. For example, in the example building, a small reduction in use of materials in the upper floor structure could reduce the overall impact by more than the whole contribution by the roof. One of the possible options to reduce environmental impact can be considered as using recycled materials. However, recycling may not always be the most environmentally friendly option. Thus, building materials containing recycled contents should be evaluated in a manner consistent with a quantitative assessment of the overall environmental impacts. Steel is the most commonly recycled building material, in large part because it can be easily separated from construction debris.

The analysis of this example building suggests that a wide range of possible outputs illustrates how useful the breakdowns are in determining which components have benefited most from choosing less environmental impact materials and which have not. Selection of sustainable building materials frequently requires difficult decisions since it is carried out considering a range of different environmental indicators. To deal with this difficulty effectively, an integrated evaluation tool, LCADesign that supports decision makers for building environmental performance has been developed and is a significant contributor to stakeholder needs in that it provides:

1. Objective detailed and comparative assessment rather than subjective assessments;

2. Real time detailed design appraisals and evaluations with tool automatic take-off CAD;

3. Generation of meaningful comprehensive graphics, tables and reports;

4. Comparing alternatives at all level of design analysis, and

5. Environmental assessment of building's development from cradle to construction.

However, environmental considerations need not be the only factor when selecting building materials. The key consideration is the material's appropriateness for the intended function as well as the cost of their manufacture; embodied energy for alternative building was much higher than for the initial example building. This is because the wool carpet was used as finished flooring to reduce the environmental impact of the scenery part in the alternative building. Even though traditional wool carpeting is non-toxic and less hazardous for building occupants, the embodied energy is higher than existing one made from petroleum products (It is also a lower cost product). While most of the environmental performance indicators do trend in the same direction for alternative products, care must betaken to ensure that the decisions are not made from a single viewpoint identified by using only one performance indicator. Thus, there will be other indicators such as costs and/or social aspects which might exist in conflict relationships when attempting to deliver the best decisions for sustainable buildings and/or building materials (Seongwon, 2002).

CHAPTER THREE

3.0   RESEARCH   METHODOLOGY

3.1   INTRODUCTION

This research work is focused at assessing the criteria for selection of building material for building construction projects within Lagos metropolis, in order to achieve this desired aim, this chapter describes the procedure and methods adopted in sourcing for data used to analyze this research work. It also discusses the sample and sampling techniques adopted, research design, research area, data for the study, data collection instruments, method of statistical analysis and statistical tools used in analyzing the obtained data.

3.2 RESEARCH DESIGN

Research design is a structure of investigating, which is focused on identifying variables and their relationship to one another. This is used in order to obtain data to enable the researcher proffer answer to the research question.

Research designs are used by researcher or investigator as a blue print for data collection prior to the actual study (Bhatt, 2010); hence it is a scheme that serves as a useful guide and platform to the researcher or investigator in his efforts to generate data for his study. Research design is used to generate primary data because when we use designs, we are involved in a direct observation of events, phenomena or an experiment. There are three main categories of research designs which include: experimental, survey and ex-post facto, but the one used for this study is the explanatory survey design.

A major feature of all survey designs is lack of control. The researcher is interested in observing what is happening to sample subjects or variables without any attempt to manipulate or control them. Explanatory survey designs are used to explain sample subjects or variables, and are focused towards collecting data to answer research questions or explain the relationship among variables, in the case of this study, explanatory design method was used.

3.3   RESEARCH ASSUMPTION

It was assumed that the professionals in Lagos Metropolis who had participated in various building construction projects are good representatives of the construction industry in Nigeria, and would be in the best position to determine how effective and efficient will the challenges / risks associated with selecting timber as a building material for building construction project within Lagos metropolis.

3.4   POPULATION OF THE STUDY

The population of this study are the professionals in the construction industry who at one time or the other have participated or had been involved in the construction of building projects within Lagos metropolis and they include the Clients, Contractors, Architects, Consultants, Manufacturers etc.

3.5   SAMPLE SIZE

A preliminary field survey would be conducted so as to help the researcher preview the sample size that will best satisfy the study, taking cognizance of the fact that the major objective of the study is to assess and identified the factors for selecting timber as a building material from the stakeholder's / professionals' perspective. These professionals include Quantity Surveyors, Engineers, Architects and Builders which were picked from Lagos Metropolis.

3.6 SAMPLING TECHNIQUE

A probabilistic sampling technique will be adopted in this research and this will be limited to the Lagos Metropolis only. This sampling technique can be referred to as the judgment sampling. According to Ogunlewe (2000), judgment sampling involves a situation whereby a researcher or an investigator is guided by what he considered as a typical case which are most likely to provide him with the requisite data or information he needs. Lagos metropolis was choosing as the reference area to be covered for this research work.

3.7   DATA COLLECTION METHOD

This section of the research comprises data collected basically by the use of questionnaires survey report form and also through personal interviews with supervisors of some ongoing building construction projects. An questionnaires survey report form was prepared for various aspects of the construction work where timber can be used as element in a building construction projects. The data used in this study are of two different sources, the primary source and the secondary source.

Primary Sources: Though restricted to Lagos metropolis only, the primary source of information for this study would be collected by the following methods:

(a) Questionnaire Method- The questionnaires are to be drawn and administered to professionals in the construction industry. The questions are to assess the choice of selecting timber as building material for construction projects.

(b) Method of Oral Interview- This will be carried out by the use of tape for vocal recording.

Secondary Sources:

The secondary data needed for this survey have been obtained from various sources after a careful search of the internet for past literature, textbooks, magazines, journals, articles and newspapers. Various research, seminar, conference and workshop papers as well as projects of other students were used in the collection of secondary data.

Although both sources are very important, the primary source of information regarded as the raw source, which will be of major interest based on the fact that the aim and major objectives is to get across to professionals in the construction industry:- the Builders, Quantity Surveyors, Engineers, Architects, and finding out their views on the choice of selecting timber as building material for construction project within Lagos metropolis.

The questionnaire would be well structured and will include both open and closed ended questions which would be used to gather responses from the selected respondents. A sample of the questionnaire would be shown in the appendix section.

3.8   DATA COLLECTION INSTRUMENT

The collection of data will be done by means of structured questionnaires and oral interviews. The questionnaire consists of three sections:

Section A deals with general information of respondents which includes location, designation of respondents, years of experience, professional qualification, etc.

Section B was designed to identify the most significant factors that contribute to the success of selecting timber as a building material in a building construction projects

Section C is aimed at assessing the criteria for selecting timber as building material in building construction projects which involve assessing the perception of various stakeholders.

To assess the factors that contribute to the success of selecting timber as a building material in a building construction projects, the survey respondents would be asked to choose an appropriate rating on the scale against each identified factor that reflects their opinions on the significant level. However, in this study, a scale of 1-5 has been used for convenience only. Most of the questions asked would require quantitative answers which involve rating using an interval scale. An example shown below identifies the level of significant with some of the factors as they would be identified by the respondent in terms of rating scale of 1 to 5 as follows;

(a)   Very Significant   5

(b)   Significant   4

(c)   Moderate   3

(d)   Slight Significant   2

(e)   Not Significant   1

The evaluations of the various factors and criteria are based on their relative importance and are subject to the data obtained from the respondents.

3.9   STATISTICAL TOOLS

The nature of the data and the objectives to be met govern the research method and the tool of research that may be adequate for data analysis. For this study, simple percentage and chi-square would be used for analysis of data.

3.10   ADMINISTRATION OF QUESTIONNAIRE

Total number of 40 questionnaires would be distributed to construction professionals in respective firms embarking in building construction projects and the administration of the questionnaire would be carried out by hand. Many follow up visits would be involved so as to enable the researcher successfully collect the duly completed questionnaires.

REFERENCES

Pai Obanya (1999): Higher Education for an Emergent Nigeria. 50t h Anniversary Lecture at the University of Ibadan.

Falayajo Wole et al.: "Assessment of Learning Achievement of Primary Four Pupils in Nigeria,September 1997.

Ngwagwu, C.C: The Environment of Crises in the Nigerian Education System. Comparative Education, vol.33, no.1 19976

Hard Lessons – Primary Schools, Community, and Social Capital in Nigeria. Technical Paper No. 420.

Adetayo JO (2008). A survey of the teaching effectiveness and attitudes of Business Studies teachers. Int. J. Labour Educ. Trade Unionism,3: 2.

Afe JO (2002). Reflection on becoming a teacher and the challenges of teacher education. Inaugural Lecture, Series 64, University of Benin.

Agbatogun OO (2006). The quality teaching personnel in Ogun State. Stud. Curr., 4: 1.

Brewer DJ (2000). High School Teacher Certification and Students Achievement. Educ. Eval. Policy Anal., 22: 129-145.

National Policy on Education (2004). Federal Republic of Nigeria,Lagos: NERDC Press.

Ferdinand JB (2007). Teachers' effectiveness and internal efficiency in primary Education. Res. Curr. Stud., 2(1): 2-6.

National Examination Council (2008). Chief Examiners Report. Minna: NECO Press.

Olatoye RA (2006) Science teacher effectiveness as a predict Students Performance in the Senior Secondary School Certificate examination.J. Educ. Stud., 6: 104-110.